Discovering Cities

Newcastle upon Tyne

*Millennium Eye Footbridge.
Photo: Michael Barke*

*Newcastle Great Park.
Illustration courtesy of Newcastle City Council*

*Tyne Bridge.
Photo: Newcastle Libraries and Information Service*

*Baltic Centre for Contemporary Art.
Photo: Baltic Centre*

*International Centre for Life.
Photo: Michael Barke*

*Former Spillers Flour Mill.
Photo: Newcastle Libraries and Information Service*

*Grey Street.
Photo: Michael Barke*

*Central Arcade.
Photo: Ray Urwin*

Discovering Cities

Newcastle upon Tyne

Michael Barke
University of Northumbria

Series Editors
Peter S. Fox and
Christopher M. Law

Discovering Cities Newcastle upon Tyne

© Newcastle Libraries and Information Service

Preface

The variety and complexity of cities as revealed in their built form has been a source of fascination to the local resident and visitor alike. Can a clear spatial structure be discerned? Why do activities cluster in distinctive quarters or zones? How do relict features throw light on the constantly evolving city?

For a long time, human geographers, regional economists, urban sociologists and local historians have sought to understand the processes which shape the city. The growth (or decline) of the city is affected by local, regional and global economic forces. The forces which shape the internal structure of the city are many and varied. There is a market in land that influences the pattern of land use and change. Public policies are often significant but can be complex to understand and difficult to follow. Social factors such as those of class and ethnic community identities are also important.

Written by urban geographers with vast knowledge and experience of the city in question, *Discovering Cities* gathers these issues together in concise and practical guides, illustrated with colour maps and photographs, to enable an enhanced perspective of cities of the British Isles.

Peter S. Fox, Chilwell Comprehensive School, Chilwell, Nottingham

Christopher M. Law, Visiting Fellow, University of Salford and Research Associate, University of Gloucestershire

© Michael Barke, 2002

This book is copyright under the Berne Convention. All rights are reserved. Apart from any fair dealing for the purpose of private study, research, criticism or review, as permitted under the Copyright, Designs and Patents Act 1988, no part of this publication may be reproduced, stored in a retrieval system, or transmitted in any form or by any means, electronic, electrical, chemical, mechanical, optical, photocopying, recording or otherwise, without the prior written permission of the copyright owner. Enquiries should be addressed to the Geographical Association. The author has licensed the Geographical Association to allow members to reproduce material for their own internal school/departmental use, provided that the author holds the copyright.

ISBN 1 84377 034 2
First published 2002
Impression number 10 9 8 7 6 5 4 3 2 1
Year 2004 2003 2002

Published by the Geographical Association, 160 Solly Street, Sheffield S1 4BF.
Website: www.geography.org.uk
E-mail: ga@geography.org.uk

The Geographical Association is a registered charity: no 313129.

The Publications Officer of the GA would be happy to hear from other potential authors who have ideas for geography books. You may contact the Officer via the GA at the address above. The views expressed in this publication are those of the author and do not necessarily represent those of the Geographical Association.

Editing: Rose Pipes
Design and typesetting: Arkima, Leeds
Cartography: Paul Coles
Printing and binding: Stanley Press, Dewsbury

Contents

Introduction	8
Historical geography	
The medieval town	11
Industrialisation and urban transformation	14
The industrial peak and suburban growth	15
The inter-war period	17
Emergence of the modern city	
The economy	20
Town planning and changing urban structure	23
Privatisation and an increasingly segmented city	26
The 'Brasilia of the North'	29
Quayside and Grainger Town: Contrasting fortunes	31
Small area studies and trails	
1. The city centre	36
2. Quayside	43
3. The Ouseburn	49
Bibliography and further information	53

Discovering Cities Newcastle upon Tyne

Introduction

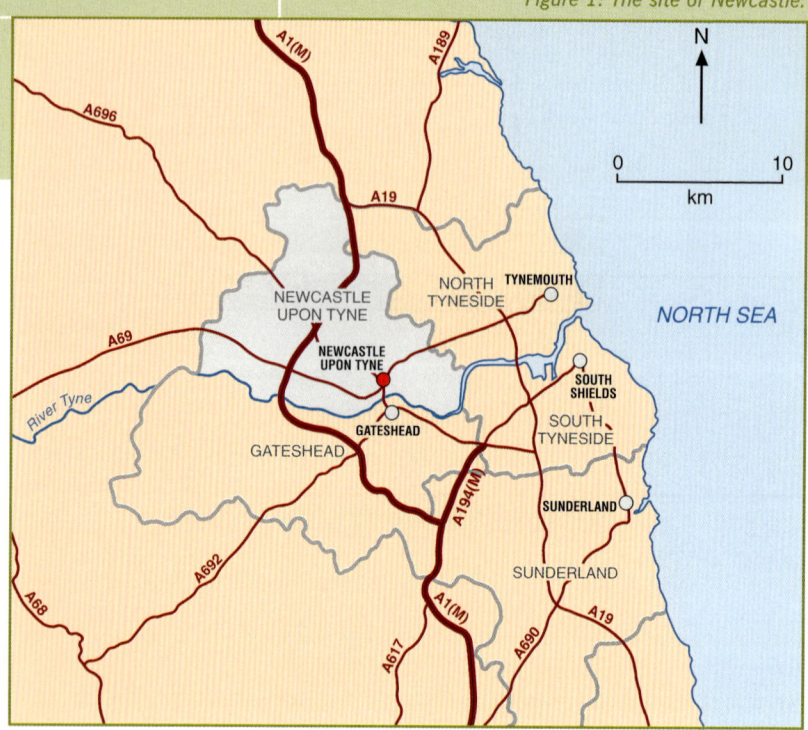

Figure 1: The site of Newcastle.

A unique and spectacular site; a distinctive contribution to national industrial history; the survival of 900 acres of open space within a five-minute walk of the city centre; two major central area restructuring schemes, separated by one and a half centuries; a lively and robust leisure and cultural scene; and some adventurous approaches to dealing with social and housing problems – all of these make Newcastle upon Tyne a distinctive and remarkable city and, for those who do not already know it, one that is well worth discovering in person.

Newcastle is, and long has been, the unquestioned capital of the North East of England. Despite its moderate population (never more than 300,000 within its boundaries), the city's excellent geographical situation has ensured pre-eminence over neighbouring areas (Figure 1).

Newcastle's distinctiveness is mainly attributable to the nature of its site, and the particular features that resulted from meeting the challenges posed by the site in terms of economic and social developments. The original settlement was located where the River Tyne cut through sandstones in the coal measures to form a steep-sided gorge and a point suitable for bridging the tidal and navigable river (Figure 2). Several small streams dissected the sandstone bluff to the north of this bridging point (roughly the site of the

Introduction

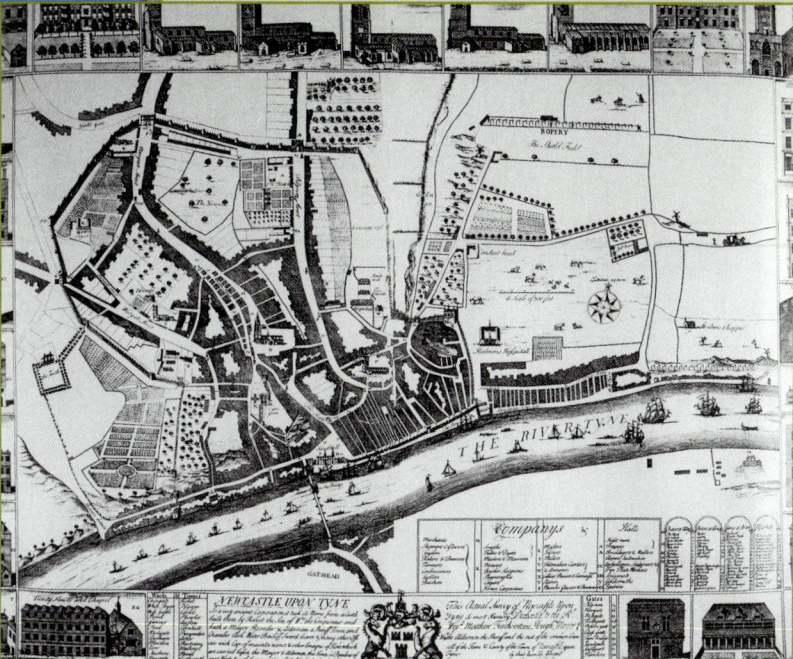

Figure 2: The site of old Newcastle.

present Swing Bridge) and created a natural defensive site commanding the river crossing. The sandstone plateau was dissected into spurs upon the most commanding of which the castle was built. Ravines created by small streams were mostly filled in during the subsequent growth of the city but are still traceable in the topography and indicated by street names such as High Bridge, Low Bridge, Stockbridge and Barras Bridge.

This site of the original settlement was approximately 18km upstream from the sea, which in modern times seems less than ideal for trading purposes. However, in earlier times an inland location on a navigable river, protected by a strongly fortified castle, provided a much more secure base for trade than one that was open to the sea. Later, in the fourteenth century, a more extensive river frontage was created on reclaimed land. By this time protection was afforded not just by the castle but also by town walls, which extended over 4km and had seven main gates and 19 towers. The security provided by the walls allowed more extensive development on the plateau, including the main markets, away from the riverside. Further natural endowments ensured the economic dominance of this site, including good quality coal seams near the surface. At a time when overland transport was hazardous and prohibitively expensive, transport on water was by far the cheapest alternative. So began the east coast coal trade and the Tyne's role in supplying London with domestic coal. Newcastle consistently and successfully sought to maintain a monopoly control over the river right through to the mid-nineteenth century. No rivals were able to challenge its pre-eminence and, to some extent, this held back more widespread economic growth within the region until the 1850s.

Historical geography

Contents

The medieval town	11
Industrialisation and urban transformation	14
The industrial peak and suburban growth	15
The inter-war period	17

Historical geography

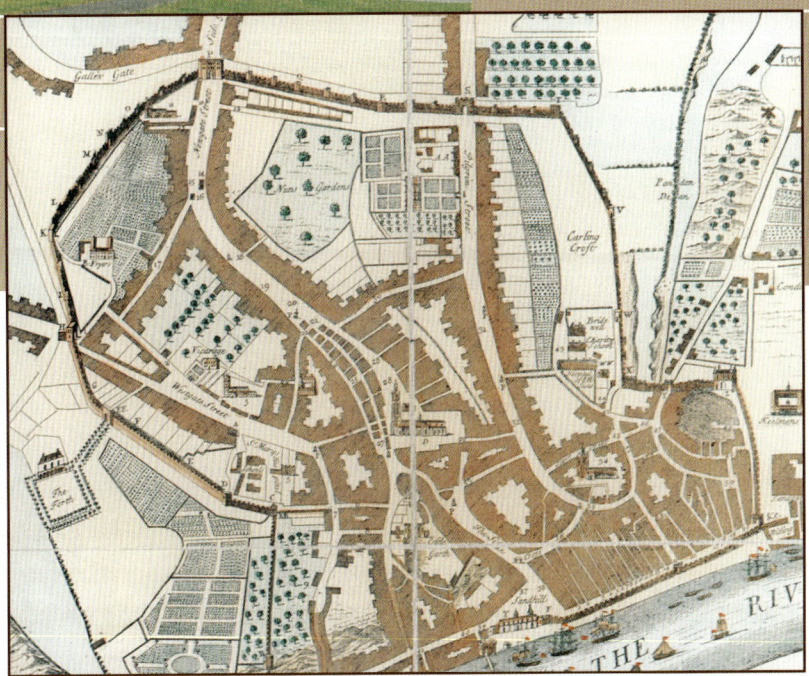

Figure 3: Newcastle in 1723.

The medieval town

In common with most other cities of regional significance, the imperatives of different phases of economic growth in Newcastle in the nineteenth and twentieth centuries swept away many of the main features of the pre-industrial city. However, more recent development has often been influenced and sometimes dictated by elements from the past.

Apart from obvious structures such as the castle, medieval Newcastle possessed three key features that still influence the way we see the city today. The first concerned street access to the pre-urban features of the bridge, the incipient riverside commercial zone in the Sandhill area, and the castle to the north. One route swept around the eastern side of Castle Hill, then swung north to form the line of the former Great North Road (now Pilgrim and Northumberland Streets). Immediately to the north of the Castle Garth, access to the main church of St Nicholas and the new markets (the present Cloth, Groat and Bigg Market area) was required and a street was created which continued to the northwest (Newgate Street and Percy Street) before turning east to join Great North Road at the present Haymarket. These two routes respectively pierced the town walls at the junctions of Northumberland Street and Pilgrim Street and Percy Street and Newgate Street and the town wall between these two later became the fixation line for the modern Blackett Street. A third route headed westwards up the high ground leading to Westgate Hill. These three main routeways heading north and west from the old core attracted linear development (Figure 3). Despite many changes in detail it is not too difficult to see how these elements form a frame for the subsequent growth of what was to become the modern city centre.

Discovering Cities Newcastle upon Tyne

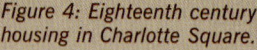

Figure 4: Eighteenth century housing in Charlotte Square.

The second feature concerned the town walls themselves. Built largely in the fourteenth century, they enclosed a much larger area than the original medieval nucleus but acted as a fixation line for much subsequent development. A number of surviving streets took their alignment from the location of these urban defences (see Trail 1). Several religious houses chose to locate at the urban fringe demarcated by the walls, and these were gradually followed by other land uses such as large residential properties. The most significant of these were the 'New House' with 'The Nuns' in the north, on a site that was to be extremely important in the early nineteenth-century development scheme for the city centre (see pages 14-15). In the eighteenth century elegant residences such as Charlotte Square (Figure 4), Hanover Square, and

Historical geography

Figure 5: Burgage plot at Old George Yard.

Clavering Place were built for the well-to-do on sites just within the walls. These were soon joined by Georgian terraced housing in streets just outside the walls, e.g. Saville Row (1770), Brandling Place and Higham Place, all seeking a more secluded environment than the increasingly overcrowded riverside area.

The third element concerned the medieval urban building units known as burgage plots (urban strip-plots held by burgesses or chief citizens of a medieval town, usually with a set annual rent which contributed to the borough's income). These consisted of elongated, linear units running back approximately at right angles to the main streets. The street frontage would usually be built up first but over time, as pressures for additional development at the urban core increased, much of the remainder of the plot would be built up, creating a characteristic line of buildings with narrow street frontages. However, where business requirements demanded it, a narrow entry from the street would open out into a linear internal courtyard. The Quayside area (see Trail 2), the Close, and especially Pilgrim Street and the Groat Market showed the most intense burgage development and building (Figure 5) and, although modern redevelopment

Discovering Cities Newcastle upon Tyne

Pilgrim Street in 1947.

meant that most of these plots were amalgamated to create larger units, their influence on urban form can still be seen in several locations (see Trail 1).

Industrialisation and urban transformation

From the medieval period to the late eighteenth century changes to the urban form of Newcastle were due mainly to natural expansion, which was later accompanied by various types of transformation of already developed areas. In the nineteenth century spectacular growth of industry and commerce took place within the town and Newcastle began to emerge as an industrial city. However, as the surrounding region also underwent industrial transformation, Newcastle became the natural location for a whole range of regional services including finance, education, legal and administrative functions, and the supply and distribution of goods.

By the early nineteenth century, the town had a wide variety of traditional industries (glass, ceramics, copperas, soap, bricks and metal working) but it was the development of engineering and its links to shipbuilding and railway locomotive manufacture that led to industrialisation on a much larger scale. Because overland transport remained difficult and expensive, a riverside location for such activities was particularly important. The rate of growth was rapid. For example, in 1817 Robert Hawthorn opened a small engineering workshop at Forth Banks employing four men; by 1840 the workforce had reached 550, and by the 1850s it exceeded 1000. Robert Stephenson opened his railway engineering workshop in nearby Forth Street (see Trail 1) in 1824 and this firm experienced similar growth. Even more dramatic was the case of William S. Armstrong who purchased land in Elswick in 1847 and built a small factory to manufacture the new hydraulic cranes he had invented, encouraged by the support of Newcastle Corporation who ordered a number for the quayside. By the turn of the century this firm employed over 20,000 people on Tyneside.

During this period of expansion a consensus began to emerge that the city required a modern central area in keeping with its increasingly dominant regional role and national significance. Several plans were proposed and discussed but it was largely due to the energy and initiative of Richard Grainger that one of Britain's most elegant and comprehensive central area development schemes came about. Grainger purchased the Anderson estate in 1834, consisting largely of 'The Nuns' and

Historical geography

Figure 6: Grey Street.

'New House' areas within the town walls (see page 12). Despite some opposition the development went ahead with the help of the sympathetic town clerk, John Clayton. Grainger's plan was a genuinely comprehensive city-centre scheme that included modern, elegant shops, public buildings, offices and arcades, and accommodation for shop owners in addition to general residential property. Three completely new streets – the elegantly curved Grey Street, considered by many authorities to be one of the finest streets in Europe (Figure 6), Grainger Street and Clayton Street – gave access to the lower town and there were several cross-connecting streets (see Trail 1). A number of architects worked on the scheme, with the major contribution coming from the locally-born John Dobson.

The completion of the new town centre gave additional impetus to the development of the railway system. In 1844 the opening of the Newcastle and Darlington Junction Railway gave a direct link from Newcastle to London, but due to the major physical obstacle of the Tyne Gorge Newcastle itself did not possess a major station. The ingenious design of Robert Stephenson's High Level Bridge solved the problem, providing railway access at a high level across the river into the upper town with a road suspended beneath it (see Trail 1). The bridge was opened in 1849 and linked directly to the magnificent new Central Station, designed by John Dobson. These developments immediately attracted commercial activity away from the riverside and created a new focus of activity. While the quayside area continued to function, a recognisably modern central business district (CBD) began to emerge in the upper town with purpose-built offices, shops and places of entertainment. As a result of these new developments, by the mid-nineteenth century a major morphological contrast was evident between the 'medieval' lower town and the 'classical' upper town.

The industrial peak and suburban growth

In the second half of the nineteenth century Newcastle changed even more dramatically, not least in terms of its

15

Discovering Cities Newcastle upon Tyne

Table 1: Population of Newcastle 1400-1991.

Year	Within official contemporary administrative boundaries	Within city boundaries as existing in 1960
1400	4000(*)	n.a.
1560	10,000(*)	n.a.
1700	18,000(*)	n.a.
1770	24,000(*)	n.a.
1801	28,294	36,194
1821	35,181	47,151
1841	70,504(a)	74,725
1861	109,108	119,185
1881	145,359	161,093
1901	215,328	249,286
1921	274,955(b)	286,533
1931	286,255	330,670
1941	291,724(c)	353,560
1951		336,400
1961	269,389	311,700
1971	222,209	
		Within city boundaries as existing in 2001 (1974 re-organisation)
1981		284,100
1991		281,700

(*) Estimates
(a) Westgate, Elswick, Jesmond, Heaton and Byker incorporated in 1835
(b) Walker, Benwell, Fenham and part of Kenton incorporated in 1904
(c) Parts of Newburn, Castle Ward and Longbenton incorporated in 1935
Sources: Middlebrook (1950); Newcastle upon Tyne Development Plan Review (1963); Population Censuses, 1801-1991.

built-up area and number of inhabitants (see Table 1). Newcastle also became the centre of one of the world's major industrial regions, shipping over 17 million tons of coal each year (of which two-thirds was exported), and producing 25% of the world tonnage of ships launched annually. Despite these massive changes, mid-nineteenth century Newcastle was, in its physical extent, no larger than the present CBD having done little more than infill the frame established much earlier. In the period up to the First World War, however, the city spread both east and west along the river, new suburbs grew to the north and the population of the central area began to decline as industrial and commercial land uses colonised existing sites.

In 1851 Newcastle had more people living within its boundaries employed in agriculture than in shipbuilding, but a surge of industrial expansion took place in the last 50 years of the nineteenth century. Much of this was due to the growth of a limited number of very large firms. The classic case was Armstrong's Elswick works in the west of the city, located on a greenfield site. In 1847 the area had a population of 3539, some 14,345 in 1861 and nearly 52,000 by 1891. Originally a manufacturer of hydraulic cranes, the firm later developed various heavy engineering products. Armstrong invented a new type of field artillery in 1855 and in 1868 the firm began manufacturing warships. It was the latter which provided the stimulus to tackle some of the problematic characteristics of the town's site and situation. The ability of vessels to sail up river to the west of the centre was

Historical geography

Figure 7: Tyneside flats, Heaton.

limited both by the river's shallowness and the nine-arched stone bridge at Newcastle. Dredging had already begun to improve access for larger sea-going ships but Armstrong's invention of the Swing Bridge (opened in 1876) allowed the west end of the city to develop even more. At the same time, although lacking a firm the size of Armstrong's, the Byker area in the east of the city grew from a population of just over 5000 in 1851 to over 10,000 in 1871 and 45,000 in 1901. Housing in both these areas consisted of terraced rows of high-density 'Tyneside flats' (Figure 7) which contrasted strongly with the large early-nineteenth century detached villas standing in their own grounds, such as Jesmond Towers and Jesmond Dene House, in the Jesmond area.

The inter-war period

At the turn of the nineteenth century, Newcastle was confident in its role as the natural centre of one of the world's premier industrial regions. However, that confidence was to be badly shaken within a generation. Although Newcastle's wide range of economic activities meant that it did not suffer as much as adjacent urban areas, the inter-war depression was real and, of course, was experienced disproportionately by the less privileged in society. Unemployment rose to over 20% (116,000 people) in 1933 and over half of those without jobs had been unemployed for more than a year.

Although the city was beset by economic problems throughout this period, a number of social advances were made. Among the most important were those in the field of housing provision. In 1921 nearly 37% of dwellings in Newcastle were of one or two rooms only and over 33% of the total population was defined as 'overcrowded', with two or more people per room. Early council estates were built under relatively generous subsidies and offered substantial two- and three-bedroom houses with gardens built in pairs or in short terraces. The layout and design of estates such as Walker, High Heaton and Pendower (Figure 8) reflected garden city principles of architecture and planning but rents were

17

Discovering Cities Newcastle upon Tyne

Figure 8: 1930s-built High Heaton estate.

way beyond what semi-skilled, unskilled and unemployed workers could afford. In the 1930s emphasis came to be placed on slum clearance with reduced subsidies for replacement housing, which led to a deterioration in space standards and building quality. Flats rather than self-contained houses predominated and the schemes tended to be located in less attractive environments than the earlier suburban general needs estates. A social hierarchy of council estates developed – a feature that remained throughout the twentieth century and helped to shape the broader social geography of the city – with slum clearance estates being perceived by many residents as much less desirable.

The local authority provision of housing in Newcastle and the rest of the North East at this time was well above the average for England and Wales. One reason for this was the city's social structure, with its relatively small middle class possessing reasonable job security and a tendency towards thrift and savings, perhaps with a building society.

Another lies with the increasing insecurity of employment for manual workers in this period. Although skilled workers were, by national standards, not badly paid, their concentration in industrial sectors, which experienced major problems in the 1920s and 1930s, discouraged involvement with the culture of security required for home ownership. Nevertheless, just over half of the houses built in Newcastle between 1920 and 1940 were built by private enterprise, though, significantly, often at higher densities than elsewhere in Britain. The typical inter-war semi-detached suburbia so characteristic of the midlands and south of England was less common in Newcastle and took rather different forms. This period also saw a substantial growth in commuting, with middle-class, owner-occupied housing being most common in neighbouring areas such as Tynemouth, Whitley Bay, Ponteland and (especially) Gosforth (not incorporated into the city until 1974).

Emergence of the modern city

Contents

The economy	20
Town planning and changing urban structure	23
Privatisation and an increasingly segmented city	26
The 'Brasilia of the North'	29
Quayside and Grainger Town: Contrasting fortunes	31

Discovering Cities Newcastle upon Tyne

Figure 9: W.D. & H.O. Wills factory.

The economy

The rearmament programme prior to the Second World War led to a brief recovery in the economy but by the late 1950s the problems that had begun to surface in the inter-war period re-emerged. It was clear that the region (and, although to a lesser extent, the city of Newcastle) relied on a narrow range of basic industries – coal, shipbuilding and heavy engineering. These industries, especially shipbuilding, became increasingly vulnerable to foreign competition, and coal began gradually to be replaced by cheap oil as a source of industrial energy. Some industrial diversification took place, for example, the W.D. & H.O. Wills cigarette factory opened in 1950 in the northeast of the city, providing a significant number of jobs for women (Figure 9). By the early 1960s, however, rising unemployment meant that the whole of Tyneside was designated to receive regional assistance. The main thrust of policy at this time was to encourage branch plants to move into the region by providing advance factories on industrial estates and various forms of financial assistance.

Table 2 shows the difference in Newcastle's structure of employment in 1960 and 1997. The 1970s was a period of particularly rapid and significant change with accelerating decline of the staple industries and determined efforts made to change the economic structure of the region and its principal city. A major shift away from manufacturing and into service industries took place with an estimated loss of 15,500 manufacturing jobs between 1961-75, leaving only about 20% of employment in manufacturing. By the late 1970s the largest employers in the city were the City Council with over 18,000 employees and the DHSS at Longbenton with 11,000. The Newcastle Area Health Authority employed nearly 10,000. In the latter part of the twentieth century, therefore, Newcastle was transformed from an industrial city to one providing

Emergence of the modern city

© MetroCentre, Gateshead

Table 2: Employment structure in Newcastle upon Tyne, 1960-97*.
Source: Newcastle City Council

	1960 Number	%	1997 Number	%
Agriculture	293	0.2	0	0
Mining, Energy & Water	8930	4.9	0	0
Manufacturing Industry	49,066	32.8	12,800	8.7
Construction	15,036	8.3	6100	4.2
Distribution, Catering & Hotels	32,251	17.9	31,000	21.1
Transport, Commerce, Banking & Finance	19,687	10.8	33,200	22.6
Public Administration & Other Services	6669	3.7	55,600	37.8
Other Services	38,412	21.3	8,300	5.7
TOTAL	170,344	100.0	147,000	100.0

(*) Note that these data relate to the official, administrative area of the city as constituted at each date. Local government reorganisation in 1974 meant that Newcastle City incorporated Gosforth and Newburn urban districts and the parishes of Woolsington, Dinnington, Brunswick, Hazelrigg and North Gosforth. In addition, changes in definitions between the two dates also make direct comparison difficult.

employment in offices, shops, education, hospitals and leisure. However, a very large proportion of these service sector jobs were in the public sector.

The late 1970s and early 1980s were particularly difficult years with the city and region being hit hard by recession. Between 1971 and 1981 Newcastle showed a 15% decline in employment. For example, Vickers, the successor of Armstrong's which had employed over 20,000 workers in the early 1900s, closed its Elswick and Scotswood works and built a new factory making tanks on the Scotswood site, employing only 300 people in 1985. But jobs were also cut in the industries that had been attracted by regional policy in the 1960s. Public expenditure cuts also meant that the service sector ceased to grow and compensate for the loss of manufacturing employment. Unemployment rates increased from under 5% in 1974 to over 20% in 1984 but they soared to well over 30% in the inner-city areas. Perhaps even more worryingly, long-term unemployment (over one year) increased within the city from 4000 in 1978 to 11,800 in 1987. This period confirmed the final dislocation of the spatial relationship that had existed between riverside industry and adjacent working class residential communities. The former all but disappeared while the latter, though much reduced, remained.

Discovering Cities Newcastle upon Tyne

Figure 10: Housing development at the site of the former shipyard, St Peter's Basin.

A number of public policy measures were taken to try and ameliorate these economic problems. For example, the inner-city areas of Newcastle along the river became a 'partnership' area with Gateshead with approved economic, environmental and social projects receiving 75% funding from the Department of the Environment. Newcastle City Council itself developed a much stronger economic development role from the mid-1970s with its creation of a specific economic development unit. In 1981 the Newcastle-Gateshead Enterprise Zone (EZ) was established, providing a reduction in planning bureaucracy, a rates 'holiday' for ten years and various tax allowances. This was fundamental in encouraging the development of the huge Metro Centre on the Gateshead side of the river. Within Newcastle, the location that benefited most from its EZ status was the riverside area in the west of the city where, as a result of the EZ incentives, Vickers built their new armaments factory. Another major development has been the Armstrong Business Park on the site of the original Armstrong works. The latter, however, was also significantly affected by the creation in 1987 of the Tyne and Wear Urban Development Corporation (UDC), a body responsible for some of the major changes in the appearance of the city's river frontage (see Trail 2 and Figure 10), and which operates independently of the local authority. The Armstrong Business Park has attracted a number of offices and call centres such as AA Insurance and British Airways, but few local residents have found employment at the site.

While the UDCs achieved much, their operation came to be increasingly criticised for achieving physical development while ignoring the needs and priorities of local communities. City Challenge was the next government-funded scheme. This sought to redress this balance by attempting to create real partnerships between public, private, voluntary and community interests, and, unlike the UDCs, local authorities had

Emergence of the modern city

Chinatown.

an important co-ordinating role. Newcastle put forward a wide-ranging programme aiming to transform the west of the city physically, and to regenerate it socially and economically. Derelict housing would be cleared and estates refurbished and modernised. The three neighbourhood shopping centres within the area that had fallen into considerable decay would be revitalised. In the extreme east of the area, adjacent to the city centre, a new Theatre Village area was to be created and Chinatown also revitalised. Industrial estates would be created and support offered for training and enterprise. The initiative ran from 1991-96 and although much was achieved the fundamental problem of high unemployment remains within the area. Over 2000 jobs were created but less than 33% of these were filled by local residents. Overall unemployment within the City Challenge area was 22.2% in 1996 compared to 10.8% for the whole of Newcastle. But some social improvement has been achieved. For example, in 1992 only 26% of the residents of Scotswood considered it to be a 'desirable place to live'. In 1995 that figure had increased to 56%. Crime rates fell between 1992 and 1995 but still remain well above the average for the city. More worryingly, demand for housing within the area has remained weak and has been a major component of a far-reaching strategy devised by the City Council to deal with the problems of the west end.

Town planning and changing urban structure

After the Second World War, town planning was a major new influence shaping the form and character of cities in the UK. House building by the public sector predominated after the war and several large council estates were built on the periphery of cities. In the late 1950s however – rather like the inter-war period – national housing policy switched in favour of slum clearance with less generous subsidies leading to poorer quality provision in the public sector. Given Newcastle's inherited stock of inadequate nineteenth-century housing, this switch was to have massive implications for the physical appearance of the city. Great emphasis was placed on building at higher densities and while Newcastle did not have the massive programme of high-rise development of some cities, flatted dwellings started to predominate, often built in 'point blocks' of five storeys, or as terraced maisonettes. The latter have been more of a problem than the much-maligned high-rise developments.

While the 1950s saw considerable development there were some in the city who were more ambitious for its future.

Discovering Cities Newcastle upon Tyne

Byker wall exterior.

The most prominent was the controversial T. Dan Smith who, from 1958, started to work towards fundamentally changing both the image and reality of the city through the medium of planning. Smith gathered around him a number of supporters and worked closely with new professional officers within the city, most importantly with Wilfred Burns the new Head of Planning. The period from the early 1960s to the mid-1970s saw the face of the city change completely, albeit in ways that not everyone approved of. Although the city centre experienced the most far-reaching plans, non-central areas were also affected, none more so than the inner east end in the Byker district. Byker contrasts markedly with the conventional semi-detached private-sector estates increasingly being built from 1970s onwards in suburban locations such as Chapel House in the west of the city.

The vision for the future of Newcastle being promoted by Smith and his supporters included a housing environment that symbolised modernity. This involved large-scale clearance of areas in the inner east and west ends. As Burns put it at the time, '... we are dealing with people who have no initiative or civic pride. The task, surely, is to break up such groupings even though the people seem to be satisfied with their miserable environment...' (Burns, 1967). Byker had been scheduled for comprehensive redevelopment from 1963. At this time it was a typical Tyneside terraced housing area, comprising Tyneside flats built at very high densities and lacking in modern facilities. In 1961 about 17,000 people lived in the area. By 1968 little had happened there apart from significant planning blight. Given its apparently poor long-term future, there was considerable reluctance to invest in Byker. Community protest resulted, to the extent that a Conservative councillor was elected in what was traditionally a staunch Labour-voting area. An international architectural consultant, Ralph Erskine, was appointed to head the scheme and he involved the community, claiming that the priorities of local residents should determine the nature of the redevelopment. Preference was expressed for new homes within Byker, the retention of key focal points such as pubs and churches, houses built in short terraces with small gardens, and play spaces for children in full view of the housing. The most spectacular aspect of the scheme – the famous Byker wall (Figure 11) – was built to provide shelter from wind but also noise from a projected motorway to the north. In the event the motorway was never built, although twenty years later the Shields

Emergence of the modern city

Figure 11: Part of the Byker wall.

Road by-pass took the same route. The scheme won much praise at the time and was hailed as a success for community-based redevelopment and for retaining members of the community in their area of origin. However, it was a relatively expensive scheme and that fact, plus its high profile, resulted in the relative neglect of other areas, especially the west end of the city. The genuine nature of the public participation involved has also been questioned and it has been suggested that, for some council officials, 'participation' provided an opportunity to persuade the public to accept the city's agenda. Furthermore, it can scarcely be claimed that the 'community' was retained in the area when it was only ever intended to re-house 9000 of an original population of

Discovering Cities Newcastle upon Tyne

Figure 12: Structure of contemporary Newcastle.

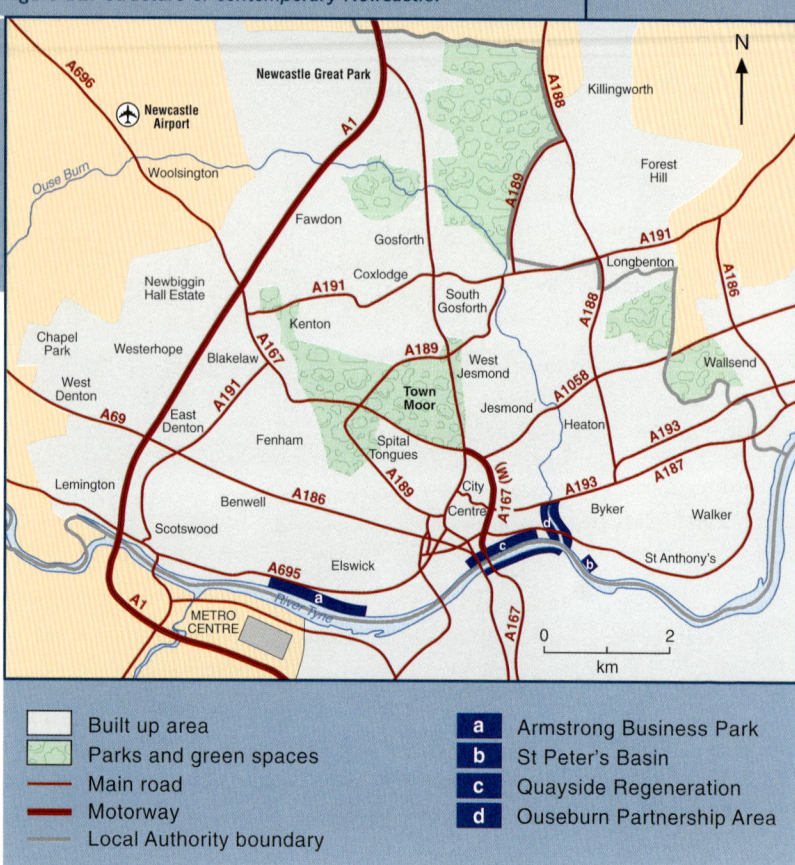

over 17,000. By the end of the scheme over 5000 households had left the area for good. In the last few years the area has shown signs of sharing some of the problems of other riverside social housing areas – low demand, high turnover and increasing social problems.

Privatisation and an increasingly segmented city

Since the 1980s there have been further important changes in the socio-spatial structure of the city, many producing considerable cause for concern. The 1980 Housing Act, which enabled council houses to be sold at heavily subsidised rates was, in Newcastle as elsewhere, very popular and about 20% of the 1980 stock had been sold by 2000. However, the spatial pattern of sales was remarkably uneven. In the four inner city wards of West City, Scotswood, Elswick and Benwell (Figure 12) only 4% of the council housing stock has been sold. One reason for this relates to the nature of the stock itself and is an important reminder of the significance of history. Predominant among the stock that was sold were conventional three-bedroom semi-detached houses in peripheral estates, especially in the north and west. Of the stock that remains, most is problematic housing – flats, maisonettes, and system-built properties. These are generally more expensive to maintain and modernise, and local authority

Emergence of the modern city

Newcastle Great Park.

housing departments lack the resources to carry out such functions. As a result of these various factors, socio-spatial segregation has increased dramatically within the city, with poorer people being even more relatively concentrated in riverside wards in public sector housing.

The Conservative government's objective of increasing owner-occupation initially produced considerably more building in that sector. Much of this owner-occupied housing has been built on 'infill' or brownfield sites. A variety of processes explain this trend. For example, while the local authority has been increasingly less able to build, it owns land, which in many cases has been sold to developers or to bodies such as Housing Associations under certain agreements. Bodies such as Health Authorities have also sought to raise money from the sale of land for private housing (Figure 13). Between 1988 and 1999, of the 5618 new dwellings built within the city no less than 88.5% have been developed on brownfield sites. In theory, such development should produce a more compact and sustainable city, with residents being easier to serve with public transport and other services, but there seems little evidence that this is the case. Although possessing the advantages of a relatively central location the residents of such areas show little inclination to reduce the number of their car journeys.

More recently, however, far-reaching proposals have been made for private sector development to the north of the city – the Newcastle Great Park. The initial proposal aroused considerable controversy and opposition due to its green belt location. Many opponents also pointed out that the development would further increase spatial disparities of wealth and lifestyles within the city. However, Newcastle City Council partially adopted the scheme (called the Northern Development Area) as part of its Unitary Development Plan and outline planning permission was granted in October 2000. The nature of the scheme has since changed, with more emphasis placed on creating employment (it is claimed that 8200 direct jobs will be created on the site) and a significant reduction in the amount of housing (117 acres accommodating 2500 dwellings rather than the 600 acres originally proposed). The City Council also successfully negotiated a scheme whereby a maximum of only 250 dwellings would be built on the site each year while private developers would have to build twice this number within the inner areas of the city.

The antithesis of the Newcastle Great Park are the riverside 'sink' estates, characterised by problems such as low demand, housing abandonment, high levels of turnover and social exclusion.

Discovering Cities Newcastle upon Tyne

Figure 13: 'In-fill' residential development by St Nicholas hospital.

The problem is most severe in the west end of the city. The City Council estimates that by 2006 there will be an over-supply of social housing – of the present stock of about 36,000 dwellings approximately 6000 will be unoccupied. Perhaps the most fundamental problem, however, is that in contrast to earlier periods there is now a whole generation, and in some areas and families perhaps two generations, of working age who have never worked. In West City ward, for example, well over 60% of children live in 'non-earning' households and are entirely reliant on state welfare. Despite the several inner city initiatives outlined earlier the problems of areas such as the inner west end have refused to go away and, as noted above, their relative difference from much of the rest of the city has become more marked. These factors encouraged Newcastle City Council to bring forward its controversial 'Going for Growth' strategy in 2000.

The strategy recognises the need to create sustainable employment opportunities if the cycle of deprivation and social exclusion is to be broken.

However, the policies to achieve this appear rather vague. Documents talk of the need to improve education and training in initiatives for local people, make much of the jobs that will be created by the Going for Growth scheme itself (presumably largely in construction and subsequently in retailing and leisure services), and discuss the need to improve transport links from the inner areas so that their residents can access jobs elsewhere. Targets are mentioned for employment generation on industrial sites but there is little information on how precisely these jobs are to be created and whether local residents will be able to successfully compete for them. Most controversy has been generated, however, by the residential and community proposals. The draft masterplan talks of 'clarifying' the community structure and is strongly reminiscent of the post-war 'neighbourhood concept', identifying the need for clear district hearts (shopping, service and leisure facilities). In the west end a larger new

Emergence of the modern city

Eldon Square.

district centre is proposed between the West Road and Adelaide Terrace in Benwell and in the east end a 'strengthened' core to the east of the existing Shields Road shopping ribbon. However, it is clear that the overall scheme involves a massive programme of demolition and rebuilding. The intention is to reduce the proportion of council housing and create sites for private sector development, mixed tenures and Housing Association activity. Angry local residents have pointed out that strong communities already exist in areas like Scotswood, but a clear implication of the strategy is that local people may be displaced in order to make sites more attractive. Some critics have even gone as far as to accuse the Council of a form of 'ethnic cleansing', and it is certainly the case that the Council made an elementary mistake in publicising their draft proposals without any initial consultation with local residents. A large-scale consultation programme is now under way but these far-reaching plans have not got off to a good start as far as local people are concerned.

The 'Brasilia of the North'

Central to the planning strategy for Newcastle from the early 1960s was the desire to restructure the urban core. Planned redevelopment of the city centre was to make Newcastle the 'Brasilia of the North' (the bold gesture of building Brazil's new capital in the interior, with modernistic architecture, had caught the general public's imagination in the early 1960s) and consolidate its position as a major regional shopping centre. The Eldon Square indoor shopping complex, the largest in Europe at the time it was built, provided 70,000 square metres of shopping floor space in two parts (Figure 14), connected by a footbridge over Blackett Street. A substantial part of old Eldon Square had to be demolished and only the western section rather incongruously remains. The southern part of the scheme links with the still thriving Grainger market of 1835. Eldon Square enabled Newcastle to compete with the vast out-of-town shopping facilities at the Metro Centre in Gateshead, and its continued vitality is indicated by the recent extension, Eldon Gardens, linked by a footbridge to the western side of Percy Street. Further development is taking place in the Newgate Street area to the southwest. Eldon Square consolidated Newcastle's chief shopping area in the north of the city centre and confirmed Northumberland Street as the main arterial retailing location, a status since enhanced with pedestrianisation. It is now almost impossible to imagine that, not too long ago, Northumberland Street was the former A1, the principal east coast route to Scotland.

Discovering Cities Newcastle upon Tyne

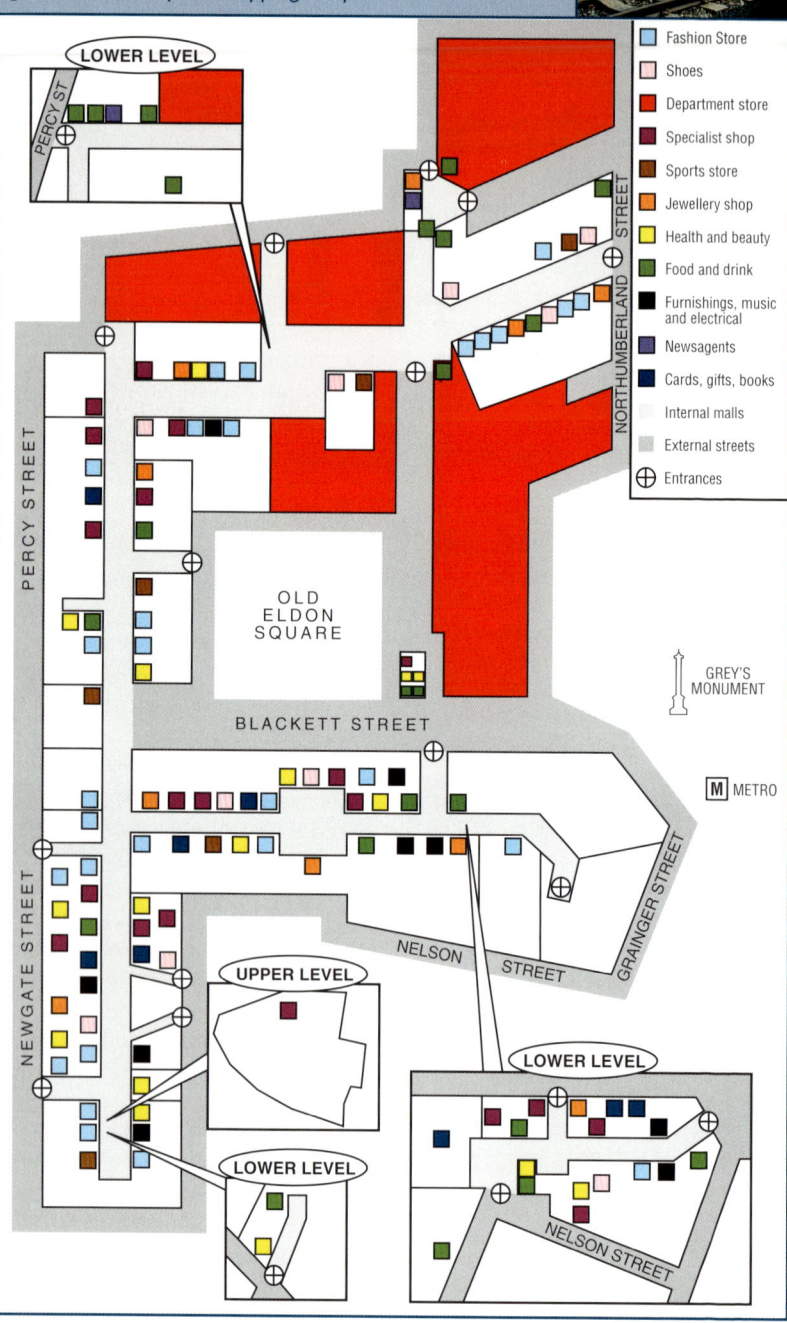

Figure 14: Eldon Square shopping complex.

Again illustrating one of the main themes of this book, namely the importance of changing relative location, the success of Eldon Square posed problems for other parts of the city centre. Former shopping streets in the southern part of the core have suffered considerably with reduced pedestrian

Emergence of the modern city

Figure 15: Pre-regeneration Quayside in the late 1970s.

flows, lower turnovers and increased vacancy rates. Significantly, Bainbridge's, one of the north's major department stores, moved from its former location on Market Street into Eldon Square, and although a neighbouring store, Binns, took over the former Bainbridge site, it closed down in 1995. The southern part of the centre, mainly the Grainger new town of the 1830s, became an extensive conservation area but there have been problems in finding appropriate uses for the buildings here. The late 1960s and 1970s saw a considerable amount of office building around the fringes of the Grainger new town but many office functions dispersed to non-central locations in Jesmond and at the Regent Centre in Gosforth. The contribution of most of this office building to the city's visual quality is, for the most part, notably undistinguished and in some cases, totally incongruous (see Trail 1).

The central area plan of 1962 included the building of a motorway 'box' with eastern and western motorways fringing the central area. In fact, the western motorway was never built and only in 2000-01 was a proper through-route completed to the west of the city centre. The final key element of the central area plan was the provision of an underground rapid transport route. This became the Metro system, which finally opened in the early 1980s and played a major role in maintaining vitality in the city centre. The main stations at Haymarket and Monument, however, have further entrenched this northern section of the central area as the most accessible and desirable location for most forms of retailing.

Quayside and Grainger town: Contrasting fortunes

If the Grainger new town gradually fell into relative decline from the 1970s, the demise of the central riverfront area of the Quayside was precipitate. Disused industrial and warehousing sites, polluted land and water and general dereliction came to dominate the area by the 1970s and 1980s (Figure 15).

The apparent success of the UDC in revitalising the Quayside area owes much to important secular changes

31

Figure 16: Quayside regeneration.

within the city and society. For example, the changing structure of employment created more jobs for white-collar professionals; changing lifestyles have reduced the significance of marriage and bringing up a family; and increasing disposable income for those in employment has greatly enhanced leisure expenditure. Added to this is the growth of Newcastle's reputation as a 'great party city' with a vibrant nightlife and, with two very large and rapidly expanding universities, a large number of single young people.

These and other factors created the demand for more office and leisure activity, which the UDC provided in east Quayside, although the rents for some of this development are heavily subsidised. Important leisure functions, including two new large hotels, restaurants and bars, and modern riverfront apartments have also been built (Figure 16). The social changes described above have also affected Newcastle's central area during the 1990s and the leisure and consumption industries have changed beyond recognition. During daytime hours the centre remains a place of work and shopping, but shopping hours have been extended and leisure activities based on eating, drinking, music and dance have grown as important parts of an 'evening and night-time economy'. The fact that cash is available 24 hours a day from ATMs assists such consumption activity. The tradition of going 'into town' on Friday and Saturday night never disappeared in Newcastle and has recently increased. Leisure services – different kinds of pubs, nightclubs and huge numbers of restaurants – have experienced massive growth within the city in the last two decades. There is increasing spatial differentiation in this leisure and nightlife activity. For example, the Bigg Market would be described by some as rather raucous and rough. Dean Street and adjacent areas have a large number of relatively cheap, standard Italian restaurants. The Quayside is slightly more sophisticated both in its pubs and restaurants. There is an identifiable, if somewhat dispersed, gay area in the south-west of the centre but the most obvious leisure concentration of all is Chinatown around Stowell Street, an area which is officially recognised and assisted by the City Council.

Emergence of the modern city

Figure 17: International Centre for Life.

However, there is also considerable variation in use of the city centre's leisure facilities at different times. The age structure of users changes significantly during the day. Such variation actually changes the sense of place of the same locality within 24 hours. At different times what for some are spaces of enjoyment and leisure are for others spaces of fear and alienation.

The object of current concern within the city centre is the area known as Grainger town. This is a designated conservation area – half of the 249 buildings are listed and 30% of these are Grade I or Grade II. Unfortunately, while attractive, most of the original buildings are often unsuitable for modern commercial uses and, increasingly, their upper floors in particular have become vacant. Also, the core area for retailing has moved to the north.

The revitalisation of Grainger town is being attempted through a partnership of Newcastle City Council, English Heritage, English Partnerships, Tyneside TEC and ONE Northeast. Public sector investment of £40 million is intended to upgrade public space and the overall environment. Various grants are also available to improve office accommodation, to assist in the conversion of vacant upper floors into housing, and to help with the repair and conservation of buildings at risk. Finance is also available for business start-ups or for established businesses in Grainger town, as is funding for individual artists and cultural organisations within the area.

Perhaps ironically, a major development lies just outside the Grainger town area but may provide the opportunity for beneficial 'spin off' into the rest of the zone. This is a major new visitor attraction, the multi-purpose International Centre for Life (Figure 17), which occupies a 10-acre site to the west of Central Station. Despite protests from those opposed to the possible misuse of genetic research, it appears that the Centre is likely to be an attraction of national significance and will provide a major opportunity for links of all kinds to develop in what had become an extremely run down part of the city centre (see Trail 1 in the following small area study trails).

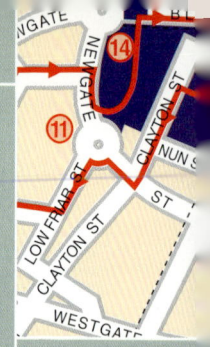

Small area studies and trails

Contents

1. The city centre	36
2. Quayside	43
3. The Ouseburn	49

Small area studies and trails

Trail 1

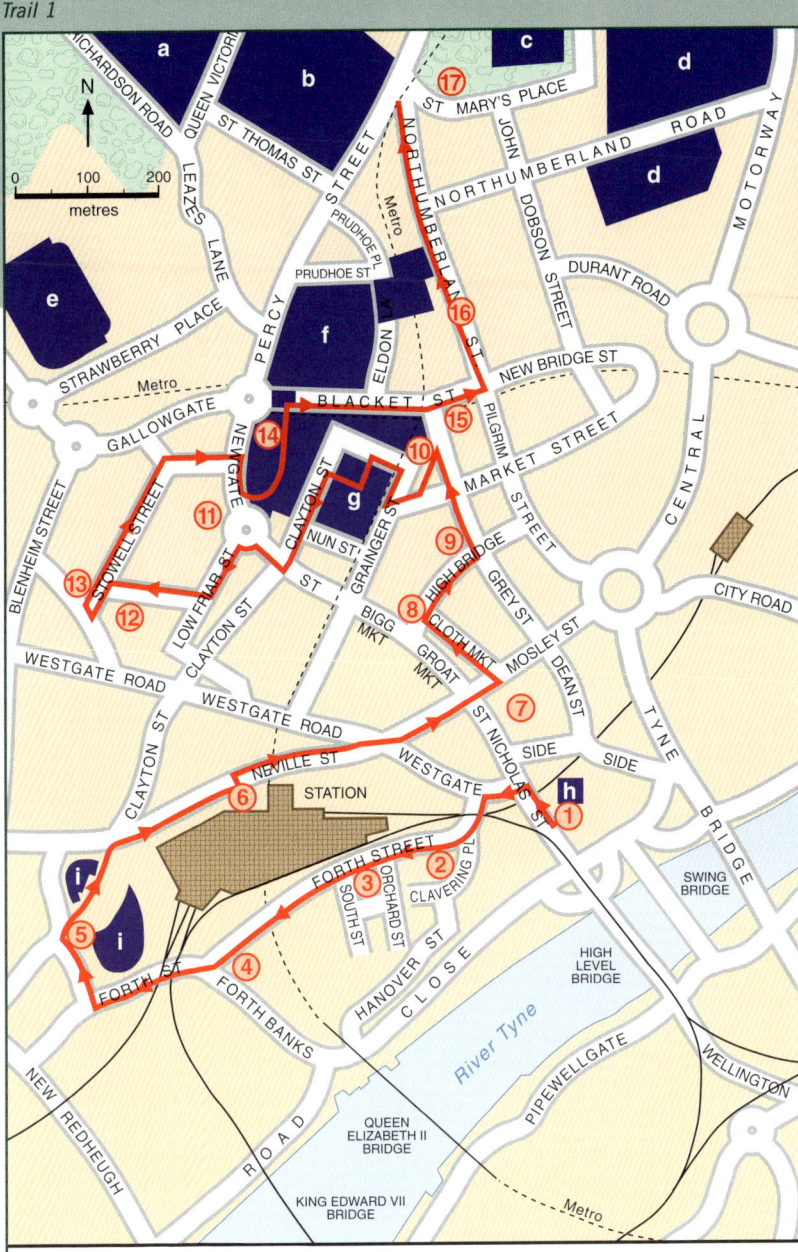

- **a** Royal Victoria Infirmary
- **b** University of Newcastle Upon Tyne
- **c** Civic Centre
- **d** University of Northumbria at Newcastle
- **e** Newcastle United FC (St James Park)
- **f** Eldon Square Shopping Centre
- **g** Grainger Market
- **h** Castle Keep
- **i** The International Centre for Life

35

Discovering Cities Newcastle upon Tyne

St Nicholas Street and Dean Street.

Trail 1: The city centre

Distance: 6km

Walking time (without stops): 1 hour 30 minutes

Disabled access: yes (although slight detours may be necessary to use lifts in Eldon Square)

This trail is concerned with two interlocking themes. First it examines the historical evolution of the urban core of Newcastle and second, it aims to demonstrate the ways in which this inherited urban form presents both challenges and opportunities for contemporary development.

1. The trail starts at the twelfth century Castle Keep, built on the site of an earlier wooden castle. The Keep is flanked on two sides by railway viaducts. The viaduct to the north was built right through the castle grounds and separated the Keep from the main gatehouse. To the southwest is the Newcastle end of Robert Stephenson's 1849 combined road and rail High Level Bridge. This gave direct access to the upper part of the town for the first time, by-passing the quayside area, and provided an important stimulus for much more intensive development north of Central Station. Walk north under the railway viaduct to the Black Gate, dating from 1247 onwards. By the early nineteenth century slum houses dominated this area and the Black Gate itself was inhabited by many poor families. The area was cleared in 1855 to make a better approach road to the High Level Bridge and the Black Gate was due for demolition but strong local protests (early conservationists?) saved it. Cross the road, walk a few yards up to Westgate Road on the left then turn sharp left underneath the railway viaduct into Forth Street.

2. On the left is the brick built Clavering House (1784) and a little further down Clavering Place is all that remains of Hanover Square. These eighteenth-century developments mark the desire of wealthier residents to escape from the riverside area to more select and quieter districts in the upper town. This peace and tranquillity was, of

Small area studies and trails

Clavering Place.

course, destroyed with the coming of the railway in 1849. The area, now the southern fringe of the modern CBD, is dominated by typical 'zone in transition' land uses such as car repairers, printers and small manufacturing workshops, some located under the railway viaduct.

3. Continue westwards along Forth Street then turn left down Orchard Street. A restored section of the town walls gives a good impression of their original height and strength. Walk along Stephenson's Lane to South Street where the remaining part of the Stephenson Locomotive Works is marked by a small plaque (although located here in 1823 the surviving building dates from 1867). The Rocket, which ran at the record speed of 58km per hour at the Rainhill trials for the Liverpool-Manchester railway, was built here. Although the contemporary urban environment is unprepossessing, the 'heritage' potential of this area was considerable. The location marks the peripheral western defences of the town and what became one of the core areas of technological innovation in the industrial revolution. However, the modern large office blocks currently being built suggest that an opportunity may have been lost.

4. Return to Forth Street and continue to the top of Forth Banks (the modern street virtually follows the line of the former Skinner Burn down to the Tyne). Another major Newcastle industrial firm, R & W Hawthorn, was located on the east side of this steep street in the 1820s, adjacent to the Stephenson factory. Quite remarkably for this difficult site, Hawthorn's also built and exported steam locomotives at this location. The firm later amalgamated to form Hawthorn-Leslie's and became a major shipbuilding concern whose main yard was in the east of the city at St Peter's (see Trail 3). Ahead to the left is the ugliest of the bridges across the Tyne, the Queen Elizabeth II Metro Bridge of 1980. To the right is the King Edward VII railway bridge, built between 1902 and 1906 to give cross-river access to Central Station from the west. This was needed partly because the High Level Bridge was becoming extremely congested, and because all north-south rail traffic pulled into Central Station from the east, with the result that the engines had to reverse direction inside the station.

5. Proceed under the railway viaduct to the west and turn right into Railway Street. On the right the new International Centre for Life occupies a 10-acre site to the west of Central Station. Ironically, this spectacular new development was built partially on the site of the former Newcastle Infirmary's graveyard (nearly 1000 skeletons were removed during construction). It consists of a £54 million complex with an interactive visitor attraction; a genetics institute (part of the Newcastle University

Discovering Cities Newcastle upon Tyne

International Centre for Life.

Medical School) which carries out human genetic research; and a Bio Science Centre whose purpose is to attract biotechnology businesses to Newcastle. The scheme was mainly funded by lottery money through the Millennium Commission but strong support came from Newcastle City Council and the Tyne and Wear Development Corporation who owned the land. It is estimated that 840 permanent jobs will be created by the development but much more important is its role as a major (national) visitor attraction, with 5700 visitors on the first weekend in May 2000 and 100,000 visitors in the first three months. Walk through Times Square. The small nineteenth century building with a clock tower is the former Cattle Market Office.

6. Turn right into Neville Street and walk in front of Central Station. Cross the zebra crossing to obtain a better view of one of the finest nineteenth-century railway stations in Britain, designed by John Dobson and built between 1845-50. The station had an important effect on Newcastle's urban structure, confirming that the commercial centre of gravity had significantly moved away from the Quayside. Adjacent buildings such as the Station Hotel and, further along in Collingwood Street, Victorian office buildings confirm this shift. Note some rather incongruous 1960s office developments, especially the massively intrusive 12-storey Westgate House that completely dominates the attractive former Union Club at 48 Westgate Road (now a Wetherspoon's pub), built in French Renaissance style. Walking along Collingwood Street it is clear that this was the main late nineteenth century office quarter of the city but equally clear that, despite the Grainger Town project, the area is having problems in finding suitable uses to occupy some of its attractive buildings.

7. At the junction of Collingwood Street and Groat Market, the Cathedral church of St Nicholas with its distinctive crown spire is ahead to the right and looking down St Nicholas Street one can see the Black Gate, the Castle and the High Level Bridge. It is now easy to appreciate how the latter gave direct access to the upper part of the town and made the market area to the left even more significant. Immediately ahead is Mosley street, again clearly part of the mid-Victorian office district. In 1880 it was the first street in Britain to be lit with electric light, developed by Joseph Swan. Between the Groat Market and the Cloth Market is the modern Sun Alliance Building built in the 1970s on the site of Newcastle's former Town Hall, itself built on the site of Newcastle's medieval market place. This was perceived as a strategic location in the mid-nineteenth century and similar concerns influenced the choice of site for the present Civic Centre at Barras

Small area studies and trails

Balmbras pub, Cloth Market.

Bridge. Looking up Groat Market note the many public houses and associated uses such as take-away-food outlets.

Pass in front of Sun Alliance House and turn left up Cloth Market. The mix of land uses is much the same, and the effects of such concentration are clearly evident if this trail is being followed on Friday or Saturday evening! On the east side of the street the buildings have a wide variety of architectural styles but their street frontages are all directly governed by the regularity of the medieval burgage plot widths. Balmbra's public house can be seen at 6-8 Cloth Market, the embarkation point of the bus to Blaydon (mentioned in the famous Tyneside song 'Blaydon Races') and formerly the site of a Music Hall. A series of lanes running along the lines of former burgage plots can be seen in this part of Cloth Market, the best being White Hart Yard with its cobbles and granite gutters, between numbers 14 and 16. Further to the north the Old George Yard, with its seventeenth-century inn, is another example. In this part of Newcastle the urban environment is more reminiscent of an English country town than a major industrial city. Most of the frontages in the Cloth Market are nineteenth or twentieth century but much of the development behind is earlier. In terms of present-day uses, however, it is clear that this area is part of the contemporary Newcastle leisure and entertainment scene, albeit at the somewhat raucous end of that spectrum. In this sense it offers something of a contrast with the Quayside area (see Trail 2). As part of the policy to occupy the upper storeys of buildings in this part of the city centre, a number of student flats have been created.

8. Turn right into High Bridge. This street takes its name from the fact that it provided access across the Lort Burn, one of Newcastle's many 'hidden streams'. The former course of the stream is marked by the dip in the lane between numbers 31 and 37. In the late 1980s High Bridge saw a short-lived reputation as a speciality clothes boutique and fashion accessory shopping street, fuelled by the growth of youth culture and, for some, increase in disposable income. Some remnants of this function remain but as with much of this southern part of the modern Central Business District the creation of sustainable commercial activity seems difficult to achieve. Walk along High Bridge to the junction with Grey Street.

9. Turn left up Grey Street. This is the core area of the Grainger-Dobson new town of the 1830s. Although the street possesses a degree of unity of style, the buildings were in fact the work of several different architects. Despite the importance of conservation the street has seen significant change; for example there

39

Discovering Cities Newcastle upon Tyne

Art deco North Eastern Co-op.

are few original shop fronts left and many of the buildings have had their interiors redeveloped leaving the facades as the only surviving authentic elements of the original buildings (e.g. the Lloyds Bank building at 102, Grey Street). Despite its architectural quality and contribution to the townscape, Grey Street is currently facing many problems in securing a long-term functional role within the city centre, as evidenced by the number of vacant premises at the southern end.

10. Walk northwards towards the Monument then turn left through the beautifully tiled Central Arcade to Market Street. Opposite this exit is the former site of Binns Department Store, the closure of which in the late 1990s caused great consternation regarding the fate of the southern part of the central retailing area. Turn right up Market Street, cross Grainger Street and enter Grainger Market via the Nelson Street entrance. This was Richard Grainger's replacement for the Corporation's butcher market that had been built in 1808 but which impeded the line of the proposed Grey Street. In 1842 there were 243 shop units in the market and Marks & Spencer's original 'penny bazaar' was located here. Much of the original detail of this classic Victorian covered market remains and the many stalls here make a huge contribution to the retail requirements of contemporary Novocastrians. Walk through Grainger market to the west side and exit onto Clayton Street.

11. Walk down Clayton Street to the junction with Newgate Street and cross the latter. Note the massive, classic 1930s art deco Co-op store on the west side of Newgate Street. The interior is even more spectacular and retains considerable store loyalty among less wealthy locals. Walk down Low Friar Street. At the time of writing, significant redevelopment (called 'The Gate') is taking place here. Pass Dispensary Lane on your right, noting the recent city centre housing conversion at Galen House.

12. Turn right under the archway into Monk Street, which continues as Friars Street. Ahead to the right is Blackfriars, surrounded by modern housing developments, part of the first attempt in the 1980s to encourage people to return to the city centre (the population of the city centre fell to less than 1000 in the 1970s). Much was made at the time of the 'risk-taking' role of the private sector in building speculative housing on such a central site. In fact, much of the site had been occupied by a disused council maintenance depot which the City made available for housing and, more significantly, the publicly funded regeneration of the historic thirteenth-century Friary as a tourism heritage site generated confidence in the area and added greatly to the attractions of the immediate environment. The housing

Small area studies and trails

Blackfriars.

has always had a very high turnover of residents but perhaps that is only to be expected, given the likely market for small, city centre apartments.

13. Turn left down Stowell Street, Newcastle's 'Chinatown'. Pass Heber Tower on the left and walk through the archway in the city walls. Down Bath Lane, to the left and right, running behind Stowell Street, are the best preserved and restored sections of the walls. The old building on the far side of the wall and ditch is Newcastle's former Fever Hospital (1804) built in an area then well-secluded from the main population. It housed the poor suffering from typhus, cholera and smallpox. Retrace your steps and turn back into Stowell Street. There is a remarkably varied collection of buildings in the street but a powerful and unified sense of place emerges from the presence of a lively and distinctive culture. The first restaurant opened in the 1960s but in the 1980s the City Council gave support and encouraged a visually distinctive area to emerge. Stowell Street is best appreciated at night or on special occasions such as Chinese New Year when the street is jammed with several thousand people of all nationalities. On the right hand side of Stowell Street about half way along is Jacobins Court (1992), another modern city centre housing scheme but one that is very much in sympathy with the adjacent Blackfriars. Turn right along St Andrew's Street to Newgate Street, with the Co-operative store on your right and the ancient parish church of St Andrews on your left.

14. Cross Newgate Street and enter the Eldon Square complex (built between 1969 and 1975) through the entrance to the right of the multi-storey car park. Take the moving staircase to first-floor level and turn left along Whitecross Way. The 'street' names within the complex attempt to impart a medieval flavour but the functional activity is very much of the twenty-first century. Like most modern indoor shopping centres, the main feeling is one of enclosure, despite the fact that for the first few years after it was built the negative reaction of many shoppers led to a significant redesign of the centre's interior to try and create an 'airier', more open atmosphere. With over 150 retail sites (not including the Green Market or Eldon Gardens) Eldon Square has managed to fight off competition from out-of-town developments, especially Gateshead's Metro Centre. Turn right along High Friars and exit from Eldon Square at Grey's Monument.

15. The top of Grey Street forms the pivot of the whole Grainger scheme and is dominated by the 135-feet tall monument, dedicated to Earl Grey for his role in the passing of the 1832 Reform Bill. There is an interesting contemporary use of space by different groups at the Monument. It is a popular location for activities such

Discovering Cities Newcastle upon Tyne

University of Northumbria, Newcastle

as demonstrations, fund raising, public speaking and live music. The nearby Metro station and one of the major entrances to Eldon Square ensures a continuous flow of pedestrians. Blackett Street, built along the line of the northern city wall, runs to the west of Grey's Monument, behind which is the wonderfully decorated Emerson Chambers (1903), now a Waterstone's bookshop, and Monument Mall, completed in 1992. The latter is a modern shopping mall on a very prominent site where any new development was going to have to respect the predominantly classical theme of the neighbouring townscape. There is considerable debate over how well this has been achieved.

16. Walk along Blackett Street and turn left into Northumberland Street. You are now in Newcastle's principal shopping street, although, in contrast with the former commercial core in Grainger Town, the buildings are generally undistinguished. The location of the popular Eldon Square shopping centre, access given by two Metro stations, extensive pedestrianisation and encouragement given to small street stalls and street performers all add to the lively atmosphere and help to explain its commercial success. However, one has to note that a number of large stores have closed in recent years (for example Henley's, Littlewoods) and the street is dominated by the outlets one finds in all city centres.

At this point you may wish to explore the northern part of the Eldon Square complex that contains some of the larger stores, most notably Bainbridge's department store, Fenwick's, which fronts Northumberland Street but also has entry into Eldon Square as does Marks & Spencer's (the latter reputedly has the second highest turnover in the UK).

17. Continue up Northumberland Street to Haymarket Metro where the trail ends. In front of you is St Thomas's church and behind this is Newcastle Civic Centre, built in the 1960s to replace the old Town Hall in Cloth Market. It is probably one of the last great manifestations of civic pride to be built in Britain. It was felt by many in the city that is Newcastle required a Town Hall in keeping with its role as regional capital, and this building was the result. But it was also intended to be the central part of the northern 'educational precinct' (consisting of Newcastle University to the west and the former Newcastle Polytechnic – now University of Northumbria – to the east), envisaged by T. Dan Smith (see page 24) as fundamental to changing the image of the city and region but also in changing its potential for self-sustaining growth. With a student population in excess of 30,000 and a massive contribution by the universities to the city and regional economy, perhaps he wasn't far wrong?

Small area studies and trails

River Tyne, looking Southeast.

Trail 2: Quayside

Distance: 3.4km
Walking time (without stops): 50 minutes
Disabled access: yes, from point 2.

Newcastle Quayside, once the principal focus of economic activity in the city, was characterised by dereliction and abandonment by the mid-twentieth century. Credit for turning around the area's fortunes is usually given to the Tyne and Wear UDC. However, as we shall see, other factors were also important. We shall also see some of the important features that remain from the past and how they are being used in the contemporary townscape.

1. Starting in front of the Castle Keep, walk towards the river down Castle Stairs. On the first landing to the right is a restored medieval well which would have been within the curtain wall of the Castle. Walk through the postern arch and turn immediately right along the footpath under the High Level Bridge to Long Stairs leading down to the Close. As is clear from some of the surviving buildings, this is the location of the residences and places of business of Newcastle merchants through to the seventeenth century.

2. The Long Stairs join the Close at the Cooperage, a barrel-making factory from the 1730s and now a popular pub, converted in 1971. Across the Close at number 35 is a modern restaurant, also occupying a group of buildings typical of pre-industrial Newcastle when access to the river, combined with storage and importing and exporting facilities, was necessary and where the wealthy merchant's residence was also his place of work. Further to the west along the Close, on the site of the former Mansion House (the earliest 'town hall'), is the modern Copthorne Hotel, built in the early 1990s and a symbol of the new role for the Quayside. Return along the

Discovering Cities Newcastle upon Tyne

Trail 2.

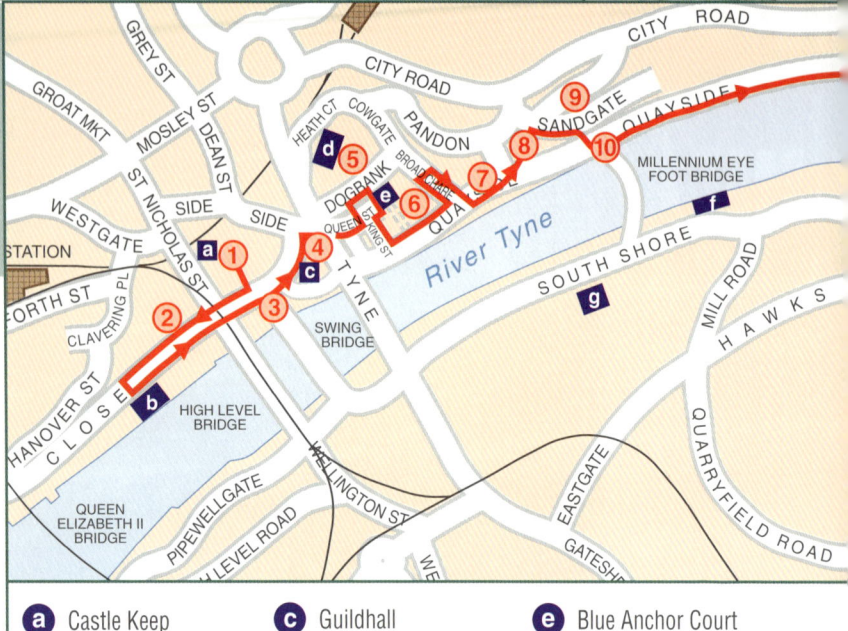

- **a** Castle Keep
- **b** Copthorne Hotel
- **c** Guildhall
- **d** All Saints Church
- **e** Blue Anchor Court
- **f** Baltic Centre for Contemporary A[rt]

Close to the east walking under the High Level Bridge and passing the old Fish Market (built in 1880) which is now used as the 'Sea' bar and night club.

3. The Swing Bridge (1868-1876) marks the site of successive crossings of the river. The original Roman Bridge was almost certainly located here, as was its successor of 1175, a stone bridge that subsequently had houses built on it. A great flood in 1771 damaged this structure and it was replaced by a new nine-arch stone bridge in 1781. However, the mid-nineteenth century dredging of the river allowed larger vessels to pass upstream and the arches of this bridge became a significant obstruction. In addition, W.G. Armstrong had developed his factory in Elswick to the west and required access to the river for launching ships and the movement of heavy engineering equipment by boat. Armstrong designed the replacement bridge to swing through an angle of 90 degrees to allow vessels to pass on either side. The mechanism was driven by hydraulic engines, also designed by Armstrong.

4. Walk along the Sandhill with the Guildhall on your right and an impressive range of seventeenth century timber-framed merchant's houses opposite. The steep street leading north into the upper town is the Side, following the line of the former Lort Burn, and the area of the Sandhill where the original port was probably located. On the east side of Sandhill, beneath the Tyne Bridge in Queen Street, King Street and Lombard Street are some impressive Victorian office blocks, built to replace timber-framed buildings which were destroyed in a fire in 1854. By the mid-twentieth century this stretch of the river ceased to be used by shipping and these offices had

Small area studies and trails

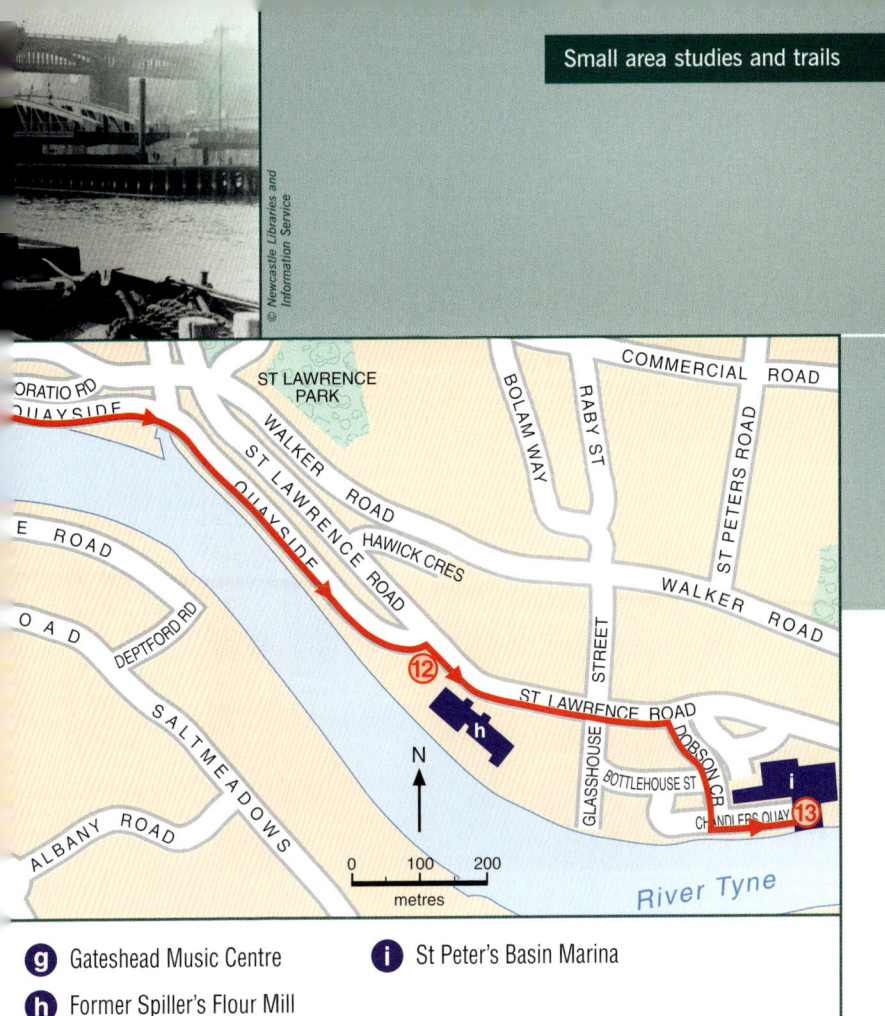

g Gateshead Music Centre
h Former Spiller's Flour Mill
i St Peter's Basin Marina

become increasingly inconvenient both in terms of their location within the city and their internal design. High levels of vacancy were recorded from the 1960s onwards. The area has changed beyond recognition in the last twenty years and this formerly empty area is now a bustling zone with restaurants, pubs, some specialised retailing, small commercial businesses and significant housing in upper floors. The area is dominated by, and gains some dramatic character from, the Tyne Bridge (1928), still the most recognised symbol of the city.

5. Walk along Queen Street to the steps leading north where there is an excellent view of the beautiful 'All Saints' church framed by buildings on each side of the stairs. At the end of Queen Street is the first modern 'new build' housing on the Quayside – Blue Anchor Quay, built in 1987 and pre-dating Urban Development Corporation activity. The size and design of the housing perhaps reflects some uncertainty over the possible success of this venture. Return to the Quayside down King Street.

6. You are now in the central Quayside area. Formerly, this area was densely occupied with buildings and many narrow alleys (or 'chares') ran back from the river frontage. The graceful Custom House of 1766 (with a façade dating from 1833) is a reminder of past commercial activity surrounded by examples of new development. The less distinguished older buildings have been demolished and replaced, or else significantly modified. The growth in leisure industries has been important in creating demand for much riverfront property. Walk along to Broad Chare and turn left past the Bonded Warehouse pub and the Live Theatre to Trinity House. This was the former headquarters of the Guild

45

Discovering Cities Newcastle upon Tyne

Crown Court, with Trinity House on left, Quayside.

of Pilots and Mariners who existed to improve the rivers, coastal waters and harbours of the north east coast for safe navigation. Immediately to the north of this area were some of the worst slums and 'fever dens' of nineteenth-century Newcastle. Although demolished, a complex pattern of land ownership in this area has led to problems in site assembly for large-scale redevelopment. Return down Broad Chare to the Quayside.

7. The new Law Courts (1984-90) dominate the area and provide an opportunity to discuss the role of the UDC in the Quayside regeneration. The need to encourage the private sector to invest in the Quayside area was central to UDC strategy and such investment has undoubtedly taken place. However, various subsidies (which are, of course, publicly financed), including important funding from European sources, have played an important role in boosting confidence. It is also the case that, in designating a special area such as a UDA, displacement effects can be triggered. Rather than generating 'new' growth, firms may move into the UDA from elsewhere in the region. For example, Newcastle's largest firm of solicitors, Dickinson Dees, moved to their office at 112 Quayside from Westgate Road in Grainger Town. It is also the case that one highly significant development pre-dated the UDC plans, namely the location of the Law Courts (publicly funded) on the river frontage. This important function represented an important statement about the future of the area and undoubtedly attracted other uses such as various legal services and restaurants. However, the Tyne and Wear UDC deserves considerable credit for making good use of the opportunities the Quayside offered, and the speed and scale of change must be admired.

8. Continue eastwards along the Quayside. Beyond the Law Courts are several brick warehouses that have been converted into luxury flats. Turn left into Milk Market. A nineteenth-century second-hand-clothes market, known as Paddy's market, was held here, and the clothes were simply laid out on the street and pavement. This was the

Small area studies and trails

Millennium Eye Footbridge.

western end of Sandgate – a notorious slum street that ran parallel to the river and became densely populated with poor Irish immigrants in the 1840s and 1850s. Its present social character is rather different. Although the UDC's social housing strategy is to have 25% of its residential development as low cost housing, the Quayside area does not contain any.

9. Walk along the new Sandgate. To the north is the Keelmen's Hospital (1701) with its prominent clock tower, built and financed by the keelmen (watermen who rowed boatloads of coal down the Tyne to be transferred to larger vessels) themselves for the poor, elderly and infirm (an early example of mutual co-operation). Most keelmen lived in crowded chares in and around Sandgate but the deepening of the river made their trade redundant in the nineteenth century. The building is now used for student accommodation. Return to the riverfront via the flight of steps that emerge opposite the Millennium Eye footbridge.

10. The area between Milk Market and Ouseburn is known as East Quayside and has seen the most spectacular change as a result of UDC activity. The UDC described the area as their flagship project, comprising a completely new business and leisure area along 1km of waterfront with, at the eastern end, about 300 housing units. About 3000 jobs have been created but as already noted, many of these are not 'new' jobs but result from relocation of firms. Dominating the area is the elegant and beautiful Gateshead Millennium Eye – a footbridge across the river leading to the Baltic Centre for Contemporary Art and the Gateshead Music Centre (currently under construction). The arts also figured prominently on the Newcastle Quayside; in a strategy to try and create a new sense of place, a considerable amount of public art was commissioned (for example the Blacksmith's Needle in front of 112 Quayside). The area provides a

47

Discovering Cities Newcastle upon Tyne

Baltic Centre for Contemporary Art.

variety of post-modern and contemporary architectural styles (The Pitcher and Piano bar being a good example). The only building to survive from the past is the former CWS warehouse, one of the first reinforced concrete buildings to be built (1899), now converted into the Malmaison Hotel. This structure in fact 'floats' on a ferro-concrete raft as the reclaimed site had marshy silt to a depth of 152 metres in places.

11. There is still considerable building activity in the eastern part of the Quayside but walking on to Horatio Street it is clear that much of this is residential. Marketed as 'chic urban living', Terence Conran-designed interiors and splendid river views ensure high prices.

12. Cross the Ouseburn and walk down St Lawrence Road. At the time of writing this is a derelict area with considerable potential. However, while the adjacent Ouseburn area (see Trail 3) has a development partnership in place, the St Lawrence quayside area lies outside its boundaries and, with the demise of the UDC, there may be problems in achieving co-ordinated development. The former Spiller's flour mill is still in use and the many industrial uses in the cheap but run-down premises of Hawick Crescent are unlikely to be compatible with the modern image of the Quayside area, even though they provide jobs for local people.

13. Walk along St Lawrence Road and cross Glasshouse Street. Ahead and to the right is the St Peter's Basin marina development, one of the earliest schemes of the UDC, built on the site of the former Hawthorn-Leslie shipyard. This development possesses an 'enclave' character, with little relationship to other local residential areas, especially the neighbouring Byker estate. An investment of £27 million produced a 110-berth marina, a few small offices and a mix of houses for sale, rent and shared ownership. The scheme has some interesting post-modern architecture. However, the area is characterised by a high turnover of residents and possesses a number of second homes and temporary lettings for important business visitors to Newcastle. It appears that no real community has developed here.

Small area studies and trails

View east from Ouseburn to Tynemouth.

Trail 3: The Ouseburn

Distance: 2.5km

Walking time (without stops): 45 minutes

Disabled access: from point 4 only.

The Ouseburn is typical of many small streams draining British cities and in its mix of recreational and industrial land uses demonstrates many of their environmental and associated problems. In the city's Unitary Development Plan (UDP) the area is designated as a 'wildlife corridor', which connects the River Tyne with open countryside to the north, and, in its southern reaches is the location of an innovative regeneration project. This trail is primarily concerned with the latter and starts at the Armstrong Bridge.

1. Looking north is Jesmond Dene, formerly part of the private grounds of W.G. Armstrong's Jesmond mansion but gifted by him to the people of the city in 1884 to use as a public park. The reasons for this generosity may not have been straightforward, however. Housing conditions in later nineteenth century Newcastle were appalling and employers were becoming increasingly concerned with the health (and therefore productivity) of their present and future workforce. The provision of public parks for recreation and fresh air was therefore seen as increasingly important for public health. More specifically, the 1870s witnessed increasing labour unrest over working hours and conditions. For the first time the large industrial employers on Tyneside had to make concessions. There was widespread concern in this group over the growing power of organised labour and it is no coincidence that, from the late 1870s, wealthy individuals made a series of charitable bequests to the people of the city – mechanics institutes, improved housing, charitable schools, lending libraries, washrooms and public open space. Armstrong Bridge, built in 1878, is now the location for a lively art and craft market every Sunday.

2. Walk across the bridge through Armstrong Park and turn south into Heaton Park, making your way down to Ouseburn Road. It is clear that, despite periodic flooding, the northeast tradition of allotment gardening still flourishes. To the right is a reminder that not all high-rise flats were necessarily bad, namely Vale House tower block, built in 1967 by the local authority. Unlike much contemporary council housing, the building is popular with residents and contains a large number of elderly households. Pass the bridge on your right, Burnville terrace on your left and take the lane to the right instead of climbing up Stratford Road. About 100m further on the Ouseburn disappears into a culvert. A steep, winding path leads you out of the valley to Stratford Grove West.

Discovering Cities Newcastle upon Tyne

Trail 3.

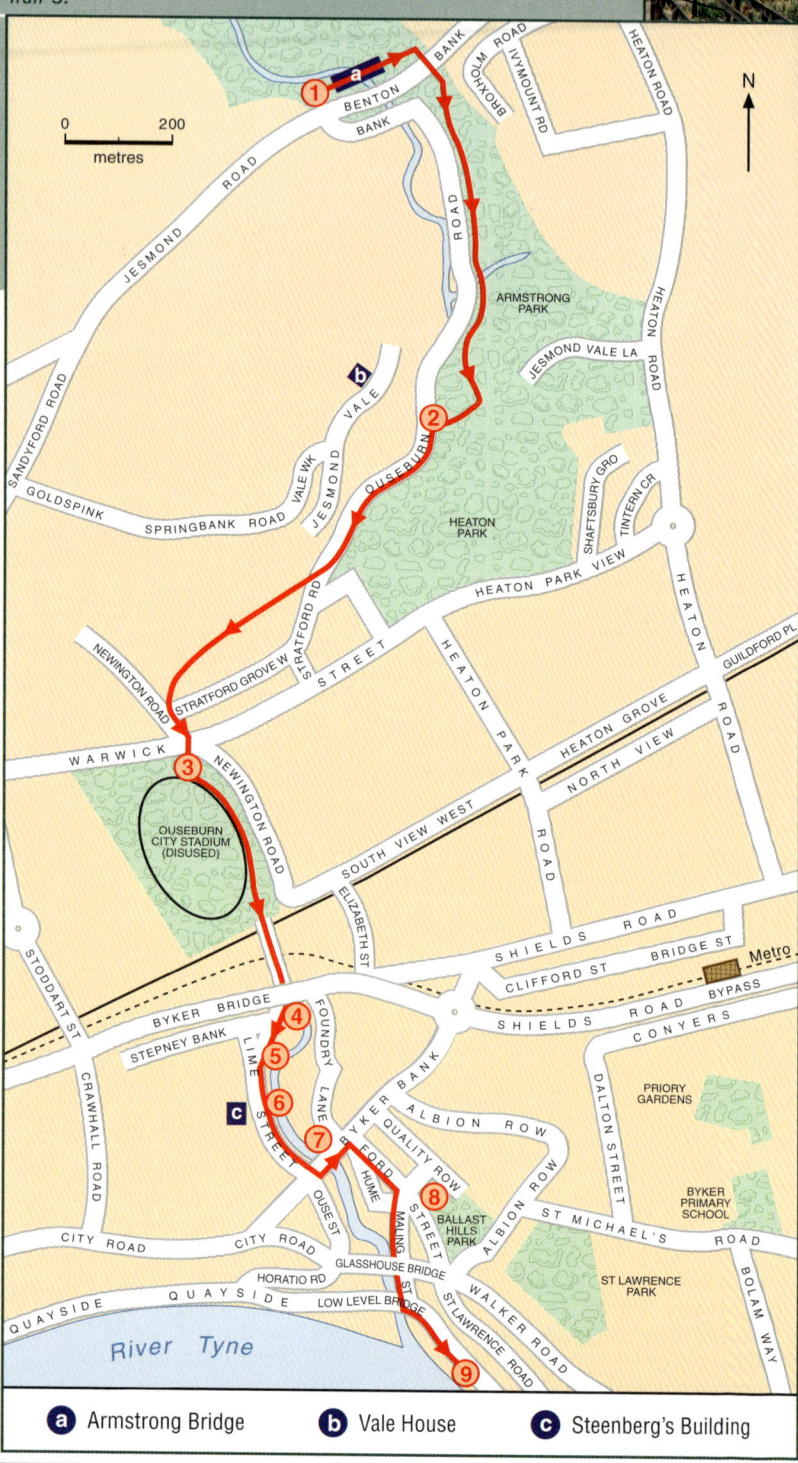

- **a** Armstrong Bridge
- **b** Vale House
- **c** Steenberg's Building

Small area studies and trails

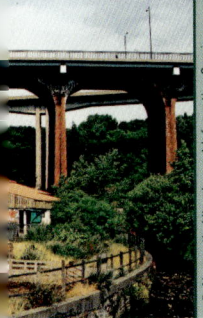

Viaduct and Byker Bridge, Ouseburn.

3. Cross Warwick Street to Ouseburn City Stadium, a public open space that would clearly benefit from some imaginative planning. At this point there is about 30-40 metres of landfill. The Ouseburn valley historically represented a substantial obstacle to communications and the eastward expansion of the town. The valley was filled in about 1906 to ease links to Byker and, beyond, Walker and Wallsend. Despite casual leisure use, the area currently lacks any real identity.

4. Walk down the steps into the Ouseburn valley and continue south under the main railway bridge (1869) and the Byker Viaduct carrying the Metro line. On the right the river emerges from its culvert. Pass under Byker Bridge (built in 1879 as a toll bridge), which connected Byker with Newcastle before the former became incorporated into the city. We are now in one of the region's first major industrial zones. Potteries, glassworks, soapworks, glue and copperas factories, roperies, foundries, leadworks and small shipbuilding yards all developed in the lower Ouseburn valley. Associated with this was a substantial residential population living in some of the worst housing and one of the most polluted and dangerous environments in Britain. Much of the development associated with this period has been demolished and a determined regeneration effort – the Ouseburn Partnership – is under way. Turn right over the small bridge over the river and pass Byker farm on your left. This is an urban farm with a broadly educational purpose, but the industrial past of the lower Ouseburn caused a major problem in 2000 when the animals had to be removed temporarily due to the discovery of lead deposits in the soil – left over from several old leadworks in the valley.

5. Walk past The Ship pub to the small triangular green area. The Ouseburn Partnership regards the area's industrial heritage as extremely important and the valley contains several listed buildings, including John Dobson's Cluny Warehouse at 36 Lime Street. Whereas not long ago the river was seen as a problem because of its pollution, it is now seen as an opportunity. The Ouseburn Partnership views the river as a unifying thread through the area and a catalyst for development. The overall intention is to create a mixed-use urban village to include residential, leisure, business and cultural uses. Although the area had become increasingly run-down, in 2000 there were still over 200 businesses of various sizes operating in the valley.

6. Walk down Lime Street. On the left is the Cluny Warehouse, which has

Ouseburn Water Sports Association HQ.

been converted into a working arts community with 50 studios or workshops of various kinds. It is planned to add a theatre, bar, café and restaurant, plus some specialist retail outlets. On the right is Steenberg's Building, scheduled to contain a heritage centre and the Ouseburn Trust offices, an indoor climbing wall and riding arena, more workshops and offices and a wide range of housing units for both rent and sale.

7. Further down Lime Street there is a good view across the Ouseburn valley. A substantial amount of light leisure craft still use the river and add to the area's atmosphere. On the far side of the river, in Foundry Lane, a number of purpose-built industrial 'starter units' can be seen and the Off Quay building which contains a further 25 studio/workshops. Turn left into Byker Bank, cross the bridge over the river and cross the road to Ford Street.

8. Some larger industrial premises are located in Ford Street and plans envisage that some of these will be removed, largely on environmental grounds. Walk along Ford Street to the corner of Maling Street, named after the famous Newcastle pottery. To your left the small green open space is the site of the former Ballast Hills burial ground where most of the poor of Newcastle were buried in the late eighteenth and early nineteenth centuries. The stone path around the area is made from the larger gravestones. Turn down Maling Street. Much of this rather nondescript area is planned to be developed for residential purposes with small-scale commercial premises incorporated in several of the buildings. Across the Ouseburn to the west, in Ouse Street, is the starting point of the 3.5km-long Victoria Tunnel, built in the 1840s to transport coal to the Tyne from a colliery inland. This is planned to become a significant heritage site.

9. Pass under the Glasshouse Bridge (1878), a replacement for an earlier crossing, and pass The Tyne pub. Cross the road to the point where the Ouseburn meets the Tyne. The new building to the east is the Ouseburn Water Sports Association headquarters, one of the key members of the regeneration partnership. Overall, the Ouseburn regeneration contrasts strongly with the Quayside area. The latter is much bigger and displays a degree of 'chic' sophistication but there is much more of a 'grass roots' feel about the former. Community activities, including performing arts, live music and various interest groups are all represented in the Partnership.

Bibliography and further information

Allsopp, B. (1967) *Historic architecture of Newcastle upon Tyne*. Newcastle upon Tyne: Oriel Press.

Barke, M. (1986) 'Newcastle/Tyneside, 1890-1980' in Gordon, G. (ed) *Regional cities in the UK, 1890-1980*. London: Harper and Row.

Barke, M. and Buswell, R.J. (eds) (1992) *Newcastle's changing map*. Newcastle upon Tyne: Newcastle upon Tyne City Libraries and Arts.

Burns, W. (1967) *Newcastle: A study of replanning at Newcastle upon Tyne*. London: Leonard Hill.

City and County of Newcastle upon Tyne (1963) *Development Plan Review, 1963*. Newcastle upon Tyne: City and County of Newcastle upon Tyne.

Colls, R. and Lancaster, W. (eds) (2001) *Newcastle: A modern history*. Chichester: Phillimore.

Conzen, M.R.G. (1962) 'The plan analysis of an English city centre' in Norborg, K. (ed) *Proceedings of the IGU symposium in urban geography, Lund 1960*. Lund: University of Lund.

Faulkner, T.F. (1990) 'The early nineteenth century planning of Newcastle upon Tyne', *Planning Perspectives*, 5, pp.149-67.

Faulkner, T.F. (1996) 'Conservation and renewal in Newcastle' in Faulkner, T.F. (ed) *Northumbrian Panorama: Studies in the history and culture of North East England*. London: Octavian Press, pp.123-48.

Flowers, A. and Histon, V. (eds) *Water under the bridges: Newcastle's twentieth century*. Newcastle upon Tyne: Tyne Bridge Publishing.

Lovie, D. (2001) 'Grainger Town, Newcastle upon Tyne, United Kingdom' in Pickard, R. (ed) *Management of historic centres*. London: Spon Press, pp. 243-73.

McCord, N. (1981) 'The making of modern Newcastle', *Archaeologia Aeliana*, Series 5, IX, pp. 333-46.

Mess, H.A. (1928) *Industrial Tyneside*. London: Ernest Benn.

Middlebrook, S. (1950) *Newcastle upon Tyne: Its growth and achievement*. Newcastle upon Tyne: Newcastle Chronicle and Journal.

Newcastle City Council (2000) *Going for Growth: A green paper*. Newcastle upon Tyne: Newcastle upon Tyne City Council.

Robinson, F. (1988) *Post-Industrial Tyneside: An economic and social survey of Tyneside in the 1980s*. Newcastle upon Tyne: Newcastle upon Tyne City Libraries and Arts.

Robinson, F., Wren, C. and Goddard, J. (1987) *Economic development policies: An evaluative study of the Newcastle metropolitan region*. Oxford: Clarendon Press.

Smith, T.D. (1970) *Dan Smith: An autobiography*. Newcastle upon Tyne: Oriel Press.

Wilkes, L. and Dodds, G. (1964) *Tyneside Classical: The Newcastle of Grainger, Dobson and Clayton*. London: John Murray.

Maps

Geographer's A-Z Map Company *A-Z street atlas Newcastle upon Tyne* (1:18,103)

OS *Landranger 88 Tyneside and Durham* (1: 50,000)

Tourist Information

Newcastle Tourist Information Centre
132 Grainger Street *or*
The Main Concourse, Central Station
Newcastle upon Tyne
Tel: 0191 277 8000

Websites

Newcastle upon Tyne City Council
www.newcastle.gov.uk

Gateshead Town Council
www.gateshead.gov.uk

Newcastle Airport
www.newcastleairport.com

Newcastle Chronicle & Journal Ltd:
www.icnewcastle.icnetwork.co.uk

Ouseburn Partnership
www.ouseburnpartnership.co.uk

Grainger Town Project
www.newcastle.gov.uk/graingertownproject

Government Office North East
www.go-ne.gov.uk

One North East (Development Agency for NE England)
www.onenortheast.co.uk

Tyne & Wear Research and Information (TWRI)
www.tyne-wear-research.gov.uk

Local Studies Library and Museums

Local Studies Centre
Third floor
Newcastle City Library
Princess Square
Newcastle upon Tyne
Tel: 0191 277 4116

International Centre for Life
Times Square
Newcastle upon Tyne
Tel: 0191 243 8200
www.centre-for-life.co.uk

Castle Keep Museum
St Nicholas Street
Newcastle upon Tyne
Tel: 0191 232 7938

Photo credits

The GA would like to thank the following for their help with photographs for this book:
- AirFotos Ltd
- Baltic Centre for Contemporary Art
- MetroCentre, Gateshead
- Michael Barke
- Newcastle City Council
- Newcastle Libraries and Information Service
- Nexus
- Ray Urwin
- University of Northumbria, Newcastle